New Wave of Inventors

Poorya Montaseri

Prepared by Ali Shabdar

ISBN: 1523796561
ISBN-13: 978-1523796564

DEDICATION

This book is dedicated to all the amazing people who always seek solutions for the world problems by ideating, tinkering, creating, and teaching others to do the same.

CONTENTS

1. INVENTION FOR SURVIVAL

The one thing that more than anything else sets us as humans apart from other species on this planet is our ability to invent and use tools. What was the one moment in history that led our ancestors to decide to pick up an object and use it as a tool to accomplish a certain task? Need or necessity, hunger or survival, all are reasons to strive towards a different path. Every day scientists are learning about and uncovering the different reasons why our ancestors from so long ago made the decision to use and invent tools, and how those decisions influenced our very evolution to where we find ourselves today.

The Oldowan – The First Stone Tools

To establish when tools first become commonly used, we have to look back several millions of years through our history. The closest living relatives to humans are the chimpanzees. They have on their own utilized tools for hunting and foraging, sharp wooden stakes like spears for hunting, specialized tools for foraging and gathering ants and insects. This suggests that our species has known how to create and use tools since humans and chimps parted ways almost four million years ago.

Next, we had what is known as the dawn of the Stone Age which, as close as scientists can place, occurred almost 2.6 million years ago in a place known as Gona, Ethiopia. These stone tools were called the Oldowan and included large fist-sized lumps of rock for hammering or crushing and the very first stone tools. These stone tools were sharp flakes of rock which were made by striking a hard stone against obsidian, quartz, flint, or any type of stone whose pieces would hold a sharp edge. Around this time scientists discovered what are possibly the first remains of butchered animal bones, with tool marks.

For around the next million years, humans didn't make any substantial leaps forward with their invention of tools. Scientists believe that it was out of necessity that our ancestors would create tools. When they wanted to butcher an animal they created a tool and then left it where it lay and so on. It is also safe to say that around this time ancient humans began searching out and utilizing caves and other convenient locations for shelter from predators and the weather. The development and use of tools occurred at a time when the planet was experiencing an intense dry period. This allowed humans to reach areas previously underwater or frozen, giving them a greater range to hunt and forage. Stone tools allowed them to take advantage of this greater range and extract the food from places they hadn't been able to venture before.

The Acheulean – Tools Advance Rapidly

When we jump ahead to approximately 1.8 million years ago, we discover that humans have taken some big leaps forward in technology as well as appearance and function. We start to discover the use and creation of small tools such as cleavers, rough knives, and even small hand axes.

Homo erectus has finally started to emerge and show characteristics that its ancestors could only dream of. They aren't hiding or climbing in trees anymore; they walk upright almost everywhere they travel now because they are now man. Now that they have moved forward in evolution they don't just start making tools to use and discard, and instead they create better, more specialized tools which are carried around, ready to use again and again. Now our ancient ancestors are using technology and invention on a daily basis; it has become an important part of their lives. Many scientists believe that this stage of human evolution is even more integral, more important than when they started using stone tools. This is where humans began to understand how important tools were to our survival and how much we needed them.

The Importance of Meat and Evolution

Homo erectus began to change rapidly once it obtained the necessary dietary food to support itself: meat. It began to develop better digestion, smaller stomachs, increased brain size, and bigger bodies. Due to the use of tools in hunting and gathering the meat after the hunt, homo erectus could now devote more dietary needs towards developing bigger brains. The meat wasn't the be all and end all, though; plants and other materials still took up a large percentage of the early human diets.

The Next Step of Society and Technology

The social evolution began to change rapidly around this time. Advances in tools meant food could be harvested; this additional food could then be stored for when it was needed. This was an entirely new concept for ancient humans who, until recently, had been gathering and using food as they could. This new social setting of keeping supplies meant women could have healthier children, which meant a better survival rate for children overall.

There were two thoughts on provisioning among our ancient ancestors. One was that the males brought food back to their partners and the other was that grandmothers were the providers for their daughters and granddaughters and so on. Apes have a very strong female group dynamic, but maybe we didn't take this from our ancient ancestors and developed our own social structure.

So what made homo erectus so good at building and creating the tools he needed to survive? What made him make the connection and achieve something that the other branches of our family tree couldn't seem to grasp? Scientists believe that early humans were much more adept at watching and recreating; we could see something being done and we could then copy it. This meant that one person who made a tool could then, just by using it, give someone else the ability to copy them and make their own tools. In this way, knowledge was spread far quicker, as one person invented the tool and everyone around them simply copied them. We were no longer working as a single unit; we were using our brains and working smarter, not harder.

There were hundreds of stone tools; everywhere people went they could pick up and use stone tools. What humans needed wasn't any more stone tools, but more ideas. We needed to create tools and invent new things, and it was in this way that we would develop and evolve.

Homo Sapiens Emerge

As we evolved further, homo sapiens began to emerge as the dominant hunter and gatherer. They began to spread out of what is known as Africa today, not only with tools but with weapons. They used the inventions they had created to build better weapons which could be used for hunting, attacking, and defending themselves from greater distances. Small blades or crude arrowheads were discovered that would have been used as tips on arrows and spears. This meant that homo sapiens could hunt prey from further distances, increasing their chances of success. They could also attack and defend from greater distances, increasing their survival chances.

A large amount of blades discovered at a location called Pinnacle Point indicates that early humans had developed the ability to heat the Silcrete to allow it to be flaked off. This was then snapped into smaller pieces, notched on one edge and blunted on the other to be attached to a spear or arrow. This wasn't any small achievement for early humans; they had to collect and process the raw materials, prepare and trim the sharp pieces, and then attach them to the weapons. By understanding these processes and then being able to pass them down through generations, this showed a much more developed brain than that of their close ancestors.

It is generally acknowledged by scientists that the process of preparation to completion of the task of making so many of these weapons would have taken a long period of time. Early humans needed to hunt and survive every minute of the day; there would most likely have been many interruptions which led to the task taking weeks or even months. This leads scientists to believe that homo sapiens had the ability and brain power to remember tasks and then come back to them for long periods of time.

Scientists believe that this combination of advanced weapons and tools combined with much more cooperative behavior by early humans was the final blow to Neanderthals. When humans first began combining these skills of learning, inventing, and teaching the use of tools and passing down knowledge, it pushed them far above their closest rivals on the food chain.

Humans now had tools to hunt, survive, and defend themselves. They had developed stronger and bigger brains and bodies, which allowed them to survive, compete with, and conquer everything around them. They had also developed a small spirit of caring for their families, which allowed them to grow larger communities and live in one place longer, which meant a community of people to help one another. When you combine these skills with their ability to teach and learn from those around them, they had begun a journey down a path that we know leads to where we are now.

2. INVENTION FOR PROGRESS

After early humans began inventing, it would inevitably lead from inventions based around their needs to survive to inventions that would lead them towards living a better and more prosperous lifestyle. When early man invented the first stone-edged blade it was a necessity because they needed this to create stronger weapons to hunt with and defend themselves from other humans and animals.

Fire is one invention that was not invented by humans, but how they controlled it, how to light it on demand, and how to utilize it was. In the early stages of human evolution, fire was an act of nature; lightning would strike a particular part of the earth that was able to ignite, and then the fire spread to surrounding areas. Early humans would often take advantage of fire to gather animals that had been killed or injured or to hunt animals fleeing from the fire. There was only one small problem with fire: early humans didn't understand it completely and, more often than not, it would turn against them.

Ancient humans used fire itself for a variety of different reasons; they used it to provide warmth, as light, to cook with, to clear vegetation, to use on the stone to create stronger tools, to heat ceramic items, and as means to scare away other people or animals. They would have most likely used fire as a means to guide people towards them like a beacon or in special social gatherings. Most early uses of fire would have been very opportunistic; it wasn't until much later that early humans began to control fire themselves.

You may not think of this as important, but shelter was a vital invention that led early humans to lead much healthier and happier lives. If you were lost in the wilderness now, your instincts would kick in, and one of the first things that you would do, would be to build yourself a shelter, somewhere to stay away from the elements and protect yourself from animals or other hostile people. But how do we know to do this? Because hundreds of thousands of years ago our ancestors discovered how to build shelters and it has been wound into us, being passed down from generation to generation.

Early shelters would have mostly consisted of caves or naturally made shelters under fallen trees or against cliff faces, etc. Next in the evolution of shelters man used stones as walls or around the base of the shelter and then layered tree branches over the top to form a roof. Structures slowly evolved using items that were plentiful in these times, such as large bones or animal skins which provided better safety and shelter from the weather. What we live in now directly links back to these very first shelters created out of stones, branches, bones, and animal hides. We have come a long way!

If it wasn't for early humans inventing clothes, we might all be walking around a lot colder right now! We probably wouldn't live as long either. Humans began to experiment and wear clothes approximately 170,000 years ago. This was around some time after the end of the last Ice Age; it may have simply been just too cold to survive otherwise, so out of necessity an invention was created. Scientists have established this timeline of discovery based on the discovery of the common lice or head lice, which were found in the remains of clothes and skins. The DNA found within the lice was tracked and used to establish a reasonably accurate timeline between humans and lice.

Changes like these in the evolution of man didn't happen over a few years either; they were spread out over hundreds of thousands of years. There was most likely a large gap between when early humans lost their hairy ape-like bodies and started wearing skins as clothes. For a while, many humans would have been running around as naked as the day they were born.

It was a necessity that drove early humans to invent items. We take a lot for granted now; almost everything that we could ever want or need is either at our fingertips or available online and delivered to our doorstep with hours. If you were taken into the wilderness, stripped completely naked, and abandoned with nothing, there is a reasonable chance that you would adapt and survive the situation. This isn't just because we are a lot smarter than our ancient relatives; we also have the advantage of thousands of years of evolution behind us. Our instincts would kick in after the initial shock and we would find or create shelter, we would hunt or gather food, and we would find a way to either make a fire to stay warm or create some form of clothing.

Inventions by early man were driven by necessity. Early man needed to eat, they needed to find some way to cook this food, and they needed a way to hunt animals much larger than themselves. You begin to establish a timeline of necessity. Early humans weren't lying around during the day with nothing better to do than invent tools and items to make their lives easier. They didn't have any free time for leisure and pleasure. They only had time for one thing: survival. They spent their time hunting for animals or gathering food in an extremely hostile environment. They had to consider other humans who would most likely kill them, large animals, and even a hostile environment. Everything they needed was extremely hard to get, and they did it all with basic tools made from wood or stone. Even gathering and constructing these tools took a lot of time and effort.

It wasn't until much later on that humans began to live lives where every minute of every day wasn't taken up simply trying to survive. As humans began to live together in larger groups, community life began to be slightly easier. Not easy by any means, but having more people around to help share the burden of survival made it much easier.

Beginning approximately 500,000 years ago, early humans began to form these larger social or family groups. Young people were being born with smaller heads and brains, which meant that it took them longer to become self-sufficient. This meant that they had to be cared for and families and groups would have most likely pooled resources to care for the young members of the group while the stronger members would hunt or gather food and resources. Around 150,000-200,000 years ago, these social groups started to develop even further. There is historical data and finds that point to groups of early humans trading and interacting with other groups up to 300 km away.

Not all of these early meetings and groups would have been friendly to each other. Many would have attacked or been attacked upon sight by other groups or individuals. Some groups may have been open to interaction, but others, like groups today, wouldn't have trusted outsiders or other groups. Survival at this stage of evolution was still the strongest instinct that pushed humans in their everyday lives.

There are two sides to this, of course. Too many people could actually make it harder to survive because of added strains on available resources, but in those hard times, the mortality rate would have been much higher. When people became sick, they died. If they were injured, they would most likely starve to death or die from those injuries. The life expectancy in early humans was certainly much, much lower than it is now. Any sick, injured, or elderly people, once they reached a certain point of becoming a drain on the economy and resources of the group or family, would have been most likely left to fend for themselves.

Once early humans had a few minutes to look around in their lives they would have begun to invent the things that started to actually make their lives easier. Instead of sitting on the ground, they may have moved a rock closer to the fire or their shelter and then used this to sit on. Instead of lighting a fire in the open, they may have enclosed it in a circle of rocks to contain it and protect the younger children. Bags for carrying items, wooden beds of branches for sleeping elevated from the cold, wet ground were all items that early humans could have lived without, but which served to improve otherwise hard lives.

Rise of the Mainstream Inventors

Historians often differ in their opinions about when certain ages began and ended. The Middle Ages were an important step forward for man when it came to inventions that not only helped man but also greatly improved our lives in general. The Middle Ages are defined by most historical sources as having run from approximately 500 AD through to 1450 AD. For a period that was often under immense times of suppression of knowledge and learning, it produced so many new inventions and ideas.

The Middle Ages were an exciting time for man in general, with some amazing discoveries and inventions that were made during a time of immense troubles.

Below are some of the most exciting inventions and when they were made, or when we can establish that they were improved on to the point that they became important. These inventions are from around the globe and, in the following sections, we will focus more on individual areas.

- Approximately 1023, the first paper money was printed in China.

- Soon after, around 1045, the first movable type of printing was invented in Bi Sheng China.

- In 1050, one of the first crossbows was invented in France.

- Around 1182, the first magnetic compass was invented.

- In approximately the year 1200, the first clothes button was invented.

- In 1202, the Hindu-Arabic numbering system was introduced to the West by an Italian mathematician, Fibonacci.

- 1249 saw Rodger Bacon invent his gunpowder formula.

- Approximately 1250, the first gun was invented in China.

- Between 1268 and 1289 saw the first eyeglasses.

- Around the year 1280, we were graced by the first mechanical clocks.

These inventions and many other inventions were coming much quicker now. No longer was it taking thousands of years to change. We could literally list thousands of different inventions from every part of the world and every manner of life. In the following sections, we will list some of the important inventions from specific regions and go into more detail.

The first regions we are going to look at are inventors and inventions from the eastern regions. Then we will look at the Persian or Iranian regions and follow that up with a general look at the Middle Eastern or Muslim-based inventions. There are thousands of small inventions, all of which are all important in their own rights, so we will try and focus on some of the bigger inventions. In the same way, not all inventions were initially made by just person or group of people alone. Often an initial idea came from many people or a person who is unknown, but it wasn't until one person took that invention to the next level that the idea really took off.

Chinese Inventions and Inventors

We can't list every invention that came out of China over the last thousand years or more, but we can give you a list of some of the most important ones. The following inventions are in no particular order. China has a long history which is full of some very creative influences, and choosing just ten inventions was a hard decision, but here you are!

Alcohol or Spirits

Nearly all of us at one stage or another throughout our lives have taken part in this little invention! The first alcohol makers known in Chinese legend were Yi Di and Du Kang of the Xia Dynasty. This was around 2000 BC – 1600 BC approximately. Research has been found showing that beer was being consumed around ancient China around 1600 BC – 1046 BC and had an alcohol percentage of roughly 4% or 5%. It was from this early beer that the Chinese discovered that, by adding additional cooked grain during the fermentation, it would boost the percentage of alcohol. Approximately around the year 1000, BC alcoholic drinks of around 11% were being created and consumed on a wider scale. This much stronger drink was mentioned by Chinese poets throughout the Zhou Dynasty from around 1050 BC – 256 BC. It took much longer for Western beer to reach anywhere near 11% alcohol. It wasn't until late in the 12th century that distilled alcohol was created in Italy.

The Mechanical Clock

We all know what this invention is and even today we all either still wear one on our wrist or take advantage of a clock as part of our smart technology, as either the phone or another app. The way we use them has changed quite a lot of the years, in the form of a wristwatch, pocket watch, digital and analog, smartphones, and on our devices or in our cars. Different variations are being used all around the world, but the first mechanical clock came from China during the Tang Dynasty around 618 – 907. Yi Xing was a Buddhist monk and also a mathematician who made the world's first initial mechanical clock. Yi Xing's clock was much larger and more complicated than what we wear today; it was a large wheel which water dripped on, forcing the wheel to turn a complete circle every 24 hours. As time went on, clocks were improved on, with different versions featuring iron or bronze, hooks, pins, locks, and even rods. During the years, 960 – 1270 a Chinese astronomer and mechanist, Su Song, built a much more sophisticated version of Yi Xing's first clock.

Tea Production and Manufacturing

Who doesn't love to sit and relax with a cup of tea? Apparently, judging by the numbers around the world, not many of us. There are thousands of different traditional teas available, as well as many different herbal varieties. Right now around the world tea is enjoying a revival, with many specific tea bars opening. According to history and Chinese legend, tea was first consumed by the Chinese Emperor Shen Nong approximately in the year 2737 BC. Later, an unknown inventor developed the small machine or device that could be used to shred the tea leaves. This was a small wooden or ceramic bowl with a sharp wheel in the center. During the years 618 – 907, tea experienced a substantial growth period. During the Tang Dynasty, tea became a favorite drink not only in China but also spread around the globe.

Silk

Man didn't technically invent silk because silkworms create silk, but it was the Chinese who invented how to gather the silk and then use it in clothes and paper. The oldest silk samples that have ever been found were dated to 3,6530 BC in the Henan Province during the Chinese Neolithic period. Silk wasn't just another item used to clothe people either. The Silk

NEW WAVE OF INVENTORS

Road was 2,000 years old and a vitally important part of Chinese cultural, commercial, and technological exchanges with the West.

Steel Smelting and Iron

Scientists have discovered that the Chinese were melting pig-iron to create iron as far back as 1050 BC – 256 BC in the Zhou Dynasty. The first truly famous metallurgist in Ancient China was Qiwu Huaiwen, who lived during the Northern Wei Dynasty. He invented the technique and process of combining wrought iron and cast iron to create steel.

Porcelain

Porcelain is a type of ceramic material which is created using extreme heat inside a kiln and is why China is named China, after all. The first types of porcelain began to appear during the Shang Dynasty, approximately 1600 BC through to 1046 BC. It wasn't until around 1708 that a German physicist named Tschirnhausen invented European porcelain, which ended the Chinese monopoly.

The Compass

It is believed that the earliest versions of the compass that the Chinese used might not have been used for navigational purposes, but to help harmonize buildings according to the ancient practices of Feng Shui. The earliest recorded history we have is that a magnetic device was used as a direction finder during the Song Dynasty 960 – 1279. The invention of the compass was a critical discovery which allowed the ship to sail much further with safety, allowing travel and trade to reach even further.

Gunpowder

Gunpowder, which was also called black powder, is a combination of potassium nitrate, sulfur, and charcoal. Gunpowder burns extremely quickly and at high temperatures which make it perfect for use in firearms and pyrotechnics. As far as most historians agree, China invented and used gunpowder during the 9th century by Chinese alchemists who were actually searching for the formula for immortality.

Movable Type Printing

The Chinese have been using a movable-type printing system utilizing wooden blocks for almost 2,000 years. During the years 618 – 907, they switched from using the rough wooden blocks to carved blocks. The next evolution was clay-style printing blocks, which marked the beginning of a major leap forward. Next they moved to metal-type blocks, but all advancements dated back to those original designs made by Bi Sheng.

Paper making

You might not think about it much, but try to go through a day without using paper or cardboard of some sort. China is considered the first country to create and refine the world's first paper. Many historians consider that the invention of paper was one of the biggest steps forward in the spread and development of human civilization. According to research, paper was first made approximately in the year 202 BC – 9 AD by Cai Lun of the Eastern Han Dynasty. This first batch of paper was made using tree bark, fishnets, bits of rope, and rags. Before they used paper to write on, the Chinese generally used carving or etching in pottery or bronze. They also used animal skins, thin bamboo strips, or silk, but all were too expensive for widespread use.

Persian or Iranian Inventions

Persia was often considered one of the cradles of science in ancient times, and Persian scientists made some major contributions to the sciences of nature, medicine, mathematics, and philosophy. Even today Iran is still striving to be known as a science hub of the world.

Qanatus

A qanat is a water management system that is used for irrigation purposes and first came from Persia. The oldest and still in use qanat is located in the Iranian city of Gonabad and is almost 2,700 years old. This qanat provides drinking water and water for irrigation purposes to almost

40,000 people.

Wind Wheels

Wind wheels were believed to have been developed by the Babylonians in approximately 1700 BC and were used to pump water for irrigation. Later during the 7th Century, Persian inventors used this initial model to develop a much more advanced wind-powered machine, called the windmill.

The Logarithm Table

During the 12th century, mathematician Muhammad Ibn Musa-Al-Khwarazmi created what was known as the Logarithm Table, developed algebra, and even expanded on the Indian and Persian arithmetic systems. Much of his work was then translated into Latin for use around the world.

Arabian or Muslim Inventions and Improvements

We owe many of the things that we use today in our everyday lives to inventions and improvements made by Muslims around the world dating back thousands of years. Some inventions weren't made by Muslims, but the basic ideas were greatly improved upon, giving us a much better idea than the original.

Algebra

Every scientist or engineer has come up with an idea or an invention, but where would they be without algebra? Not very far, that's where! We all use algebra in our lives every day, maybe not directly, but many of the things we use wouldn't be here if it weren't for algebra. It wasn't until around the 12th Century that a British scientist or scholar Robert of Chester translated Al-Khwarizmi work.

Toothbrush

Every time you brush your teeth in the morning or night, then you have Islam to thank for your clean teeth. The Islamic religion stresses the importance of good hygiene and, although it may not be responsible for the first toothbrush (Egyptians used a twig from the toothbrush tree known as a mistake), they are definitely responsible for the spread of cleanliness and good health.

Marching Bands

Some of the very first marching bands that we enjoy all across the world date all the way back to the Ottoman Mehterhane. They weren't the same as the marching bands that we experience in sports and parades today; these original marching bands started playing at the start of the battle and wouldn't stop playing until either the army won or retreated. It was during the wars between Europe and the Ottoman Empire that these military bands made such a big impression that European armies adopted the idea for their own armies.

Guitar

Next time you turn on MTV or watch your favorite band play during a concert, when you hear the guitar you owe it to ancient Arabic culture. The modern guitar began from the Arabic Oud, which is a lute with a bent neck. During the Middle Ages, it made its way across to Muslim Spain, where it was called a qitara. The guitar that we all know and use today has many influences, but the biggest influence was the Arabic lute.

The Magnifying Glass or Glasses

The Arab world didn't just revolutionize the way we look at mathematics, but they also revolutionized the way we see things. An Arab scholar Alhazen Abu al-Hasan from Basra was one of the first people to describe in detail how the eye works. He conducted many experiments with different reflective materials that proved to the world that the eye didn't use "sight rays" to sense the environment. It was during these experiments that he also discovered that you could use curved glass as a source of magnification. The first magnifying glasses were reading stones produced by Alhazen; it was these first reading stones that eyeglasses were based on.

Coffee Drink

They may not have invented coffee, but they helped to spread its popularity all around the world. It is believed that coffee originally came from Ethiopia, but it didn't gain popularity until it spread around the Arab world. It was during the 16th and 17th centuries that coffee was introduced to the European markets. It was in approximately the 17th century that the first coffee was brought to the market of London. Venice had its first coffeehouse open in 1645, and Germany had its first real taste of coffee during the Turkish retreat from Austria in 1683. Historians state that, during their retreat, the Sultan's soldiers abandoned hundreds of weapons, equipment, and sacks of coffee.

Hospitals

The hospitals that we all use and take advantage of around the world are all based upon a basic template that was first created in Cairo. In the year 872, the Ahmed Ibn Tulun hospital was established and provided free health care to all patients. An essential idea of hospitals was used before this, but it was this hospital in Cairo that we used as a template to base our future hospitals on.

It isn't surprising to learn that many of the things we use today in our everyday lives originally came from ideas many years ago. What is surprising to learn is which items led to the items today and where their ancestors came from.

Many of the different things that we take for granted in our lives all originated somewhere, and it is an interesting journey to see where they came from, what they started as, and what they have all evolved into. One person or a group of people can come up with an idea, but until that plan is perfected or worked on, that original idea is going nowhere. It can take many years for an initial invention to be perfected and really create an impact where it can begin to spread and gain momentum around the world.

A good example of this timetable of evolution is the clock, which started as a large, cumbersome device made from a wooden wheel which relied on water dripping to turn the wheel over a 24-hour period. If you have ever looked inside the mechanism of a wristwatch or clock you understand just how complicated they really are. There is a reason that when your watch breaks you take it to a specialist for them to look at. Inside are hundreds of cogs, gears, and mechanisms all working together in perfect synchronization to tell you the time.

Now we have taken watches to the next level; we have digital clocks, wristwatches, and, everywhere throughout our daily lives, cars, phones, TV, etc. It is interesting to look back at where the first clock came from and where it is now; now our watches actually have all the technological benefits of our smartphones.

Western Inventors

As we have covered Eastern inventors and their inventions, we are also going to write about Western inventors and some of their more popular and world-changing inventions. We could write pages and pages about all the different inventions and how they impacted a wide variety of different people, but we just don't have the room! So what we will do is focus on some of the most important inventions and how they impacted the world around them and generations into the future.

To make it easier to read, we will break it down into several different countries located within Europe.

British Inventors and Their Inventions

Precision Screw-Cutting Lathe

It might not sound like the world's most exciting machine, but imagine the possibilities that this machine unlocked for manufacturing around the world. We take a lot for granted when we walk into a hardware store and grab a box of 1,000 screws all identical for a few dollars. In 1797, Henry Maudslay created a precision machine that allowed him to mass produce screws of the same size. Until this machine was invented almost all screws were made by hand, but by mass producing screws it unlocked many more manufacturing possibilities.

The Hovercraft

The hovercraft was invented during the second half of the 20th century by Christopher Cockerell. He made his first full-size model in 1959. Capable of carrying four men, it crossed the English Channel as part of its maiden voyage. This first full-size model was known as the Saunders-Roe Nautical One or SR-N1. Hovercraft technology allows the craft to function over water, wetlands, swamps, and small patches of dry land. This technology has been utilized for both military and commercial craft around the globe.

The Television

It may have come a long way from where it started, but the general idea is still the same and probably one of the most popular inventions we have seen: the television. John Baird was a Scottish engineer and is considered to be one of the inventors of the television. He produced a live, moving, greyscale image on a television screen using reflected light. This discovery was made in his room in London in 1925, and he then went on to demonstrate the first color transmission in 1928. The BBC used his 30-line Baird system from 1929 to 1932 before they switched to a rival system.

The Jet Engine

War is a powerful driving force when it comes to inventions. It is often under extreme pressures that man finds his greatest achievements. Frank Whittle was an RAF pilot who managed to create and market a jet engine to the Air Force at the outbreak of World War II. His jet engine technology was placed into the Gloster E.28/39, and later into the production model the Gloster Meteor. He later went on to receive a knighthood for his service to king and country.

The Marine Chronometer

If war isn't a good enough motivator for inventions, then offering rewards for solutions to problems is definitely a good way to speed up the process. The British government offered a huge £20,000 bonus to whoever could invent a machine that would establish a ship's longitudinal position at sea. John Harrison finished his first marine chronometer in 1735, but it wasn't until 1759 that he completed the winning design.

The Bouncing Bomb

Barnes Wallis invented a round-shaped bomb that was used in the 'Dam Busters' raid on the Ruhr Valley. His inspiration for this bomb was the cannonball and how sailors used to skip the cannonball across the surface of the waves to increase its range. With little to no funding and even less optimism that this could be accomplished, Wallis ran the tests, did the calculations, and concluded that, with a seven-degree angle, it would work.

After a successful test on an abandoned dam, the English successfully destroyed both the Mohne and Eder dams, successfully damaging the German's war efforts.

The Reflective Cat's Eye

This is one little invention that may seem like an obvious one and one we take for granted, but once upon a time we drove around at night with just lights on our cars and no really effective way to see lines on the road. Percy Shaw got the idea for the reflective cat's eye stud after driving home one night having trouble seeing and then noticed how the light from his car reflected off a cat's eye. Sales were slow to begin with, but after gaining approval from the ministry and blackouts during the war which limited the use of lights dramatically, sales went through the roof. Sometimes it is the smallest idea that can lead to the most popular inventions.

French Inventions and Their Inventors

The Hair Dryer

A French hairstylist, Alexandre Godefroy, was the brains and inspiration behind the first hair dryer in the year 1888. This was a far cry from the models that we all use today; it wasn't small or portable in any way. It was a larger model which was used inside the salon alone, but the luxury of finally having a way to dry hair was worth the size and the inconvenience.

Canned Goods

Sometimes, unfortunately, inventors come up with ideas that change the world and influence everyone's lives but get a little reward for it. This isn't the case with Nicolas Appert, who became a very wealthy man with his invention. This invention was another one that was motivated not only by necessity but by a reward. In 1795, Napoleon offered a reward of 12,000 francs to anyone who could come up with a cheap and easy way to preserve a large amount of food. This invention was another invention motivated by war because Napoleon needed to feed his vast army while they were marching. Appert took out the prize with a glass jar design that was then sealed with wax.

Braille

Louis Braille was a young man who was blinded in both of his eyes while still a child. As a gifted young man, he was accepted into France's Royal Institute for Blind Youth. It was here that he began work on what we know today as Braille. Braille is an efficient way for people suffering from blindness to be able to read and write. His invention, the system of Braille, wasn't acknowledged as anything worthy until well after his death; his valuable contribution at the time wasn't appreciated, and he died knowing that he had created something great, but he didn't receive any praise for it.

Mayonnaise

We all appreciate a great sandwich, and it wouldn't be the same without mayonnaise. It is the small things that we all take for granted that we would miss if they were suddenly unavailable. Mayonnaise was created to celebrate a victory when other sauce ingredients weren't available, and it was named after Port Mahon, where it was conceived when Duke de Richelieu won a victory over the Spanish.

The Hot Air Balloon

Two brothers were the brains behind the hot air balloon. Joseph-Michel and Jaques-Etienne Montgolfier successfully launched their first unmanned hot air balloon on the 10th of September, 1783. It was in November that same year that they successfully launched an untethered and manned hot air balloon flight.

This is just a sample of two countries from Europe; there are many other European countries and many other inventors throughout Europe. Some of these we will go into more detail on in the rest of this chapter, but, unfortunately, there are just too many to list individually.

This isn't to say that these other inventors weren't just as important or more so, only that there are only so many different inventors that can fit in one book!

Post-Renaissance Inventions and Inventors

The Renaissance period ranged from AD 1300 through to AD 1600 and marked a period of awakening in humanity. It was a period where mankind moved from out of a not so bright Middle Ages into a brighter time of discovery and knowledge. There was a large amount of positive discoveries throughout the fields of science, technology, art, politics, faith, and religion. All of these changes led us as people towards a much brighter, happier, and socially better time in our lives. There were a lot of different discoveries during this period, so we will focus on some of the most important ones and try to cover as many as possible.

The Steam Engine

The actual idea of an engine powered by steam was first introduced by a Greek mathematician who was known as the 'Hero of Alexandria.' Basic steam engines were developed during the sixteenth century.

Thomas Savery was one of the first developers, utilizing a steam engine to power a water pump. This is considered one of the first steam engines ever invented. Savery constructed his first steam engine in 1698, and it was known as a 'fire engine.' He made a gift of his invention to the Royal Society of London. He dedicated his life to the steam engine, going before King William III and also securing a patent for his information.

This first steam engine required a large amount of coal to produce the steam, which meant it wasn't as economical as desired. Further changes to Savery's design were introduced by Thomas Newcomen, who created a better version, calling it an 'atmospheric engine.' James Watt made another modification, attaching a separate condenser which reduced the amount of coal needed by almost 75%. This breakthrough made steam engines much more economical to use.

The Printing Press

It is recognized by many historians that it was Johannes Guttenberg, a German goldsmith, who developed the first printing press. He borrowed money to begin it in AD 1336, and it was completed in AD 1440. The printing press that Johannes developed utilized metal letters which could be quickly and easily interchanged with other letters. The letters were replaced by lines; ink applied to the plate and paper pressed against the plate to create the printed impression.

To make a practical demonstration of his printing press, Johannes printed copies of the Holy Bible. The Bible at the time contained 42 lines per page, and some copies of this printed Bible are still available today. The invention and introduction of this printing press revolutionized the way knowledge was spread: quickly, easily, and much more available to everybody.

The Telescope

The invention of the telescope is believed by most historians to have been made by Galileo Galilei. Galilei heard that a Dutch lens maker was offering an instrument that could be used to see distant objects. He then set out to create his own magnification device. It was in 1609 that he began using this device to observe objects in the sky, becoming the first person to do so.

His first attempt only managed to have a magnification power of three, but with hard work, he raised this to a magnification level of thirty. He then used this to study the moon in detail, learned that Jupiter had four moons, and that the Milky Way was made up of stars.

Mechanical Clock

As we discussed previously, man began using different forms of clocks as far back as 4000 BC. The first clocks were extremely primitive, made of wood, stone, and utilized water or the sun to determine the time. The clock then gradually took different forms and over time began to take on different forms utilizing different technology. Some of the first mechanical clocks used mercury and drums, the flow of mercury passing through holes in drums and being stored in small compartments. The introduction of these mechanical clocks allowed time to be measured as twenty-four hours and also in fractions.

It is believed by most historians that it was Filippo Brunelleschi from Florence, Italy who invented the mechanical clock. Leonardo Da Vinci contributed a lot to the development of the clock but doesn't hold any credit for its invention.

Artillery and the Cannon

During the second, third, and fourth Mysore wars rockets were utilized to good effect. Some rockets were recovered and transported to England, where William Congreve led the mission to develop even better missiles. In the year 1805 AD, the Royal Arsenal, the British armed force's research and development center of the time, demonstrated solid rockets by test firing them. Congreve utilized firing tubes to boost the accuracy of the rockets.

The Microscope

Two inventors, Hans Janssen and Zacharias Janssen, both have the credit for inventing the microscope. Both father and son were respected spectacle makers and had a passion and interest with experimenting using lenses. They were believed to have developed the first compound lens in 1590, with Hans creating it and his son Zacharias developing the idea further.

It was later during the 17th century that Anthony Van Leeuwenhoek developed microscopes with a magnification power of up to 270 times the original size of the item.

Flush Toilets

We all appreciate a good flush toilet! It wouldn't be much fun to have to go back to throwing a bucket out of our window every morning! Humans, in general, are believed to have been using flush toilets as far back as ancient BC, but the credit for introducing the idea of a flushing water closet of the Renaissance period goes to John Harrington.

He built his first flushing water closet in Kelston, England. This design featured a valve which let water out of a tank and then washed through the toilet, emptying the bowl. This design interested Queen Elizabeth I and she chose a similar design to be built in the Royal Palace. The design worked well with one problem: the ventilation of the system was inadequate, allowing the sewer smells to leak back into the Royal Palace. This was overcome by the placement of herbs and fragrances around the rooms inside the palace.

Matches

The modern matches we all utilize today were first designed by Robert Boyle, the first person to produce fire using the chemical reaction of two chemicals. Through many experiments, Boyle discovered that by rubbing Phosphorus and Sulfur together, it would create an instant flame reaction.

As a result of these discoveries and further experiments in 1827, the modern match was developed by John Walker, an English chemist. He utilized Antimony Sulfide and Potassium Chloride, gum, and starch to create his fusion matches.

The Inventions of Leonardo da Vinci

The actual term "Renaissance Man" hails from the fifteenth-century Italy and refers to the idea of a person who has knowledge and skills a wide variety of different areas. The one person who holds more in common with a true renaissance man is without a doubt Leonardo da Vinci. He was known as a scientist, architect, artist, inventor, and engineer.

You may know Leonardo da Vinci as a famous artist, but he actually spent more time working on his technology and science endeavors. His ability to draw and sketch his ideas in detail worked in his favor when it came to his ideas and inventions. His sketchbooks show many ideas that da Vinci had well before they were ever invented, many years later. A lot of these inventions da Vinci sketched well before the technology existed actually to dream of creating them.

Leonardo da Vinci didn't just stick to one particular theme when it came to his ideas, sketches, and creations. He drew and designed weapons for war, water systems, flying machines, and even tools. He was known for his ability to look beyond what was considered traditional; that is what made him such an excellent inventor.

Da Vinci noted time and time again that he hated not only war but the idea of killing his fellow humans. The trouble was, like many inventors, da Vinci needed money to support his research and household and found that it was easier to convince wealthy people to support him with war machines. These war machines were paraded as a way to easily conquer and kill his wealthy patrons' enemies; perhaps it is lucky for all of us that none of these were ever made. The world could have turned out a much different place.

In the following section, we will go through in more detail some of Leonardo da Vinci's more creative and popular ideas and inventions.

The Ball Bearing

Not that exciting? Think again. Consider how many machines utilize the ball bearing as a necessary part of their design. The original idea of the ball bearing is believed to have come from as far back as the Roman Empire, but historians believe it was da Vinci who drew the first practical designs. Many of the devices that da Vinci came up with not only feature them but also rely on them heavily.

The Parachute

You wouldn't want to jump out of a plane without one! Da Vinci was fascinated not only just with flight but also human flight more specifically. His pyramid-shaped design was covered with cloth, and he wrote that it would allow people "to throw himself down from any great height without suffering any injury." Recreations of his design in the 21st century suggested that his design would work just as da Vinci believed it would.

The Machine Gun

Da Vinci didn't name his invention the 'machine gun' and instead he chose '33-barrelled organ.' It wasn't exactly a machine gun; it didn't fire multiple bullets out of one barrel, but it could deliver a massive volley of force relatively quickly. Had someone decided to build his machine, they would have quickly decimated any approaching ranks of infantry, possibly changing how wars were fought.

His idea was to mount eleven muskets side by side on a plank or board, then mount three of these together in a triangle pattern. A plank could then be mounted through the middle, allowing a shaft to be placed to spin the entire setup. Every time one set of eleven guns was fired, the next could be loaded while the third cooled down.

3. INDUSTRIAL REVOLUTION

The Industrial Revolution took place from the 18th through to the 19th centuries, and it was a period during which predominantly rural and agrarian societies throughout Europe and America became urban and industrial. The Industrial Revolution began in Britain in the late 1700s, but before this most manufacturing was accomplished in people's homes, using hard work and basic tools.

Industrialization marked a specific push away from cottage industries and towards powered, specific, special-purpose machines, factories, and mass production. It was the textile and iron roles, along with the development of economic and reliable steam engines, that played the most pivotal roles in the Industrial Revolution. It was these improvements that led the way towards many other improvements in society. Due to increased production and urbanization, society needed better systems for transportation, better communication systems, and better banking systems. Every change, every effect, led to another change and had a flow-on effect on every aspect of people's lives.

Without the Industrial Revolution, there is no doubt that the lives we are living now would be extremely different, but it wasn't all good news for everybody. Unfortunately, life wasn't great for everyone during this period; it often meant people from lower classes ended up in almost slave-like employment, poor living conditions, and bad health, which resulted in shortened life spans.

The Birthplace of the Industrial Revolution, Britain

Before the machines, before factories, most people lived in small rural villages. Their daily life would revolve around the village life, animals, or farming. People didn't go to work to make a wage and then buy the goods that they needed. They raised their animals, farmed their lands, and produced almost every single item that they needed to survive. This wasn't always the easiest of lives and led to little opportunity for people to rise above the position that they were born in.

There were a few different factors that all contributed to Britain becoming the home of the Industrial Revolution. The first and biggest reason was that Britain had huge reserves of both coal and iron ore, both essential items for industrialization. Britain itself also had a politically stable environment, as well as being one of the world's largest colonial powers. This meant that Britain could call on its colonies to provide not only raw materials but also labor, and it had a marketplace and infrastructure in place for the goods that it produced.

As the demand for the goods Britain was producing increased, the merchants creating these items needed to come up with more cost-effective and time-effective ways of producing them. This is what led to the factory system and the rise of mechanization.

Industrialization Through Innovation and Invention

The textile industry was one of the first industries transformed by industrialization. Before the era of industrialization, most textiles were created in people's homes. Merchants would normally provide basic equipment and any raw materials and then pick up the finished product. This system required the workers to regulate their own hours and output, which made determining results difficult.

In 1764, an Englishman, James Hargreaves, invented what was known as the 'spinning jenny' or just Jenny. This allowed an individual to produce some different spools of thread at the same time. By the time he died in 1778, there were believed to be as many as 20,000 spinning jennies in operation all over Britain. Another inventor, Samuel Compton, improved upon the spinning jenny and in approximately the 1780s an English inventor, Edmund Cartwright, invented the power loom. This allowed the cloth to be woven by machines instead of by hand for the first time.

It was during this period that another Englishman, Abraham Darby, discovered a much more affordable way to produce cast iron. Darby developed a furnace that used a coke-fueled system instead of the more common charcoal-fired systems. In 1850, a British engineer, Henry Bessemer, created a cheap and easy way to mass produce steel. Iron and steel were the main materials used to create tools, machines, ships, buildings, and the infrastructure needed to maintain the Industrial Revolution.

Another important part of the Industrial Revolution was the steam engine. In 1712, Thomas Newcomen developed what is considered one of the first practical steam engines. This initial model's main purpose was to pump excess water out of mines to prevent flooding. Around the year 1770, a Scottish inventor named James Watt improved on Newman's original version. The steam engine later went on to power a wide variety of machinery, trains, and locomotives and ships during the Industrial Revolution.

Transportation During the Industrial Revolution

During the Industrial Revolution, it was also proven necessary to upgrade the way goods were moved around the world. Before machines and mass produced goods, most items were moved around the countryside by horses and wagons or hauled up and down rivers and canals on boats. At the beginning of the 1800s an American, Robert Fulton, constructed the first commercially successful steamboat. By the middle of the 19th century, there were steamships carrying goods back and forth across the Atlantic Ocean.

The same time that steamships were beginning to forge their way across the world, the steam locomotive began to come into its own. It was in 1830 that the Liverpool and Manchester Railway became the first to offer a timetabled passenger service to the public. By the year 1850, Britain was proud to boast over 6,000 miles of railroad track.

During the year 1820, a Scottish engineer, John McAdam, created a new way to build and create roads. It was his technique, known as macadam, that resulted in roads becoming easier to travel on, smoother, and longer lasting. It also meant that during the months of winter, goods could be moved from the countryside much easier.

Banking and Commerce During the Industrial Revolution

It was the necessity of being able to communicate during the Industrial Revolution that drove inventors to create quicker and easier ways to communicate. Two British inventors, William Cooke and Charles Wheatstone, patented the first commercial electric telegraph in 1837. By the year 1840, all railways utilized the Cooke-Wheatstone system. Later in 1866, there was a telegraph successfully laid across the Atlantic Ocean, allowing communication between countries to be much quicker than ever before.

With all of these new inventions impacting the way people lived, traveled, and worked, it was never going to be long before the way they used money and banked changed too. A stock exchange was established in London during the 1770s, and the New York Stock Exchange in the 1790s. It is believed by many historians that it was a Scottish man, Adam Smith, who was responsible for founding modern economics.

Living During the Industrial Revolution

The Industrial Revolution made mass produced goods much more available to everybody and greatly improved the living conditions for middle- to upper-class people. Unfortunately, life for the lower class in many cases got considerably worse.

The wages and working conditions for those working in factories was low and dangerous. To make up for low wages, people would often be forced to work extremely long hours in dangerous positions. The job positions were monotonous and unskilled, which meant that people could be easily replaced and this left very little job security.

Children also formed part of the workforce during the Industrial Revolution; as much as one-fifth of the workers were children. They didn't just do the easy jobs either. Unfortunately, because of their small size, they were often employed to work in confined spaces or cleaning machinery.

As the urbanization of once-rural areas spread during the Industrial Revolution, the influx of people couldn't be accommodated. This crowding into bad conditions meant that the housing was inadequate, along with almost non-existent services. Disease and pollution spread quickly throughout crowded slum areas, and fires often rampaged out of control for days. It wasn't until the later part of the 19th century that living and working conditions improved with the introduction of labor reforms introduced by the government. It was also around this time that workers had the opportunity to begin forming their own trade unions.

The Spread of the Industrial Revolution

Britain tried to stop the spread of the Industrial Revolution to other countries. They also went as far as enacting legislation to try and prohibit the export of both their technology and the skilled workers supporting it. Their plans to stop the spread of industrialization inevitably failed, and it wasn't long before countries like Belgium, France, and Germany all had industrialized. The next leap for industrialization was across the Atlantic Ocean to the United States. By the middle of the 19th century industrialization was established in the Western part of Europe and the northeastern part of the United States of America. By the beginning of the 20th century, it had spread across much of the world and America was the world's foremost industrial nation.

Drop in Mainstream Inventions

Inventions fueled the beginning of the Industrial Revolution and also the industrialization of the Western world. This rapid advancement of humans also fueled a massive increase in the human population. Not only did we now need more of everything—clothes, food, medicine, and homes—we also needed a better quality of these things.

As an example of how quickly the population grew, just before the beginning of the 21st century, the population was at approximately 6 billion people, almost a 400% increase in the population in only one century. In the 250 years from when the Industrial Revolution began, the human population increased by almost 6 billion people.

This growth in population is tied together with some natural resources that we use and also the amount of pollution that we produce.

Before the Industrial Revolution humans were living in a sustainable way, consuming fewer resources than were being produced. After the Industrial Revolution, humanity moved into an area of using more resources than we could sustain, at the end affecting the way we lived and the conditions that we lived in.

Unfortunately, the entire Industrial Revolution and the growth periods that came after it relied heavily on fossil fuels for energy. Fossil fuels by their very nature are quickly used and slow to be replaced. The fossil fuel era will be one of the shortest eras in humanity's time; almost as soon as it began it began to decline only several hundred years later, and other sources of energy would need to be developed as soon as possible.

Modern Inventions and the 20th Century

As we move into the 20th century, we see an increase in the speed of inventions and innovations such that we have never experienced before. More growth was experienced in the 20th century than any other century in the past. As an example of this, we just need to look at what was invented during this time.

We started the beginning of the 20th century with things like extremely basic airplanes, automobiles, and early stages of radio. These inventions were amazing and excited humans, providing ideas and experiences that we had never imagined before. Next, we look at where we ended up at the end of the 20th century only one hundred years later.

By the end of the 20th century, we have spaceships flying into outer space, we have computers, cell phones, and the internet. When you consider how long it took the early man to learn how to uses the tool, humans have come a long way in a very short time!

Inventions During the 20th Century

There are a lot of inventions that happened during the 20th century; we will try and cover some of the more important and popular inventions. In a small amount of time, there were some major advancements and inventions. Unfortunately, we could fill more than an entire book with them! In only 100 years humanity made some fantastic leaps forward, some for the better of humankind and others for its detriment. The following inventions are just a few of them!

Nuclear Power

Nuclear power was considered a game changer when it came to providing a source of power during the 20th century. We found ourselves going from a dirty and polluting power source, coal, to what seems like a pollution-free, efficient, and seemingly unlimited power source,

Sadly for man, nuclear energy turned out to be a double-edged sword. Not only could it be used to create an energy source, but it could also be used to create weapons powerful enough to destroy all of humanity. While nuclear power plants didn't throw pollution straight into the atmosphere like coal power plants, when they failed it ended up destroying entire regions. This happened at Chernobyl in 1986 and also again in Fukushima in 2011 when a tsunami and earthquake cause a spectacular meltdown with global ramifications.

Overall nuclear energy when used properly is a sustainable energy source with very little pollution. It will only remain to be seen if we can use it correctly and responsibly into the future.

Personal Computers

As we sit here and type on our laptops or read this on our tablet or desktop PC, it is hard to image where we would be without computers in our life. Computers began to be used as far back as World War II, but I don't think that you would like that computer anywhere in your home, if it would even fit to begin with! The Apple computer was first introduced in 1976 by Steve Wozniak and Stephen Jobs, and it changed the world! Since that first introduction, even Apple has had its up and downs, but now almost everybody throughout the world has their own computer.

They are continually changing, evolving, and being recreated and have changed so much from where they began that if you sat both models next to each other, they would be almost unrecognizable.

The Airplane

The train may have made a huge difference in the 19th century, but the airplane opened up an entire world of possibilities in the 20th century. Nowadays you can fly anywhere in the world within 24 hours. It may take longer depending on flights and airports, etc., but if you had your own plane, pilot, and runways I'm sure that you could do it easily! Planes didn't just make it easier for us to travel around either. We use planes to transport goods, fight fires, dust crops, storm watching, and much more.

Like most inventions, planes can also be used for good or bad, depending on how you look at them. Planes have revolutionized the way we fight wars; no longer do you look a man in the eye, or even across a field. Now we can drop bombs dropped within inches from almost outer space.

The Automobile

Although the automobile was developed at the end of the 19th century, it never really became a practical or popular means of transportation available to everybody until the 20th century. It was Henry Ford's assembly line that truly made the automobile reach new heights of popularity. Automobiles made it possible to transport goods across the country quickly, reducing transportation time and allowing perishable goods to be moved via truck. First cars were only available for the extremely wealthy, but that changed quickly, and soon roads were being created all over the world.

Rockets

Rockets may have been invented thousands of years ago by the Chinese and used throughout history ever since, but it wasn't until the 20th century that man really managed to control them. In the 20th century, rockets became not only much bigger, but they also became more powerful and much more controllable. Now, we have something that was intended as a weapon of war suddenly being utilized for other reasons, like sending a man into space or placing satellites into orbit.

The Submarine

Like many inventions of the 20th century, the submarine was actually invented earlier, but the true innovations came in the 20th century. It started to gain popularity during service in World War I; it was seen as dangerous, but it really didn't play any major role during the first war. All that changed when Germany unleashed its submarines during World War II. Submarines were responsible for sinking more tons of shipping than any other means during World War II.

When countries began combining nuclear power with submarines, they created ships capable of running for almost unlimited amounts of time and distance, also producing very little sound or any emissions. Submarines now make up the main parts of many navies throughout the world; given any situation submarines can be moved without knowledge to any sea throughout the world at any time, making them a truly deadly adversary.

The Antibiotic

It wasn't until Alexander Fleming discovered penicillin in 1928 that we had any chance to prevent and beat every bug that came along. Penicillin was just the first in a wide range of antibiotics that came along and helped to combat and fight diseases. Image the plague sweeping through our planet nowadays if we had no antibiotics to take—it would decimate us.

The Radio

As we drive around in our cars listening to music on the radio, it is hard to imagine the impact that the radio had on us in the 20th century. The first major accomplishment of the radio was allowing people to be heard all around the world, without a cable or wire of any sort. Before the television came along, families gathered around the radio to listen to the news and shows.

The Television

It may be filled with a lot of junk now, but imagine where we would all be without the television. We used to tell stories around fires, write them, or draw them on walls and eventually write everything down and pass them on through books; now we have television. Now we have the television broadcasting news and entertainment around the globe almost the second that it happens 24/7, 365 days a year.

The Internet

When the computer came along it basically ended the typewriter overnight; sure, some people still use them, but realistically they are obsolete and time-consuming. Computers are great, but without the internet, the computer would be basically gathering dust in most people's homes. When used correctly the internet can be used to find out almost anything, from any time and in any part of the world! We can't all visit the great pyramids of Egypt, but we can look and learn about every part of them on the internet, whenever or wherever we like.

4. THE NEW DAWN – 21ST CENTURY

Invention or innovation: the 21st century has been a significant century when it comes to inventions. Listing them all is going to take more time than we have. Every year that goes by seems to offer us more inventions and new innovations.

History has shown us, and we have discussed in previous chapters, just how curious and smart humans really are. From many years ago when the first caveman came out of his cave and struck two stones together to create a sharp-edged tool all the way through to solar power, electricity, and the Internet.

We will try and cover some of the more popular and exciting inventions and some of the 21st century inventions that are have been shaping human evolution. If we miss one of your favorite inventions then we apologize; there are just so many to choose from. Not all of these inventions are purely life or death; some are purely superficial, but important nonetheless.

The following inventions are listed in no particular order.

The iPod and iTunes from Apple

The iPod and iTunes took the way we all listened to music and turned it on its head. The geniuses at Apple Inc. gave mankind a new way to collect and listen to their music wherever in the world they happened to find themselves. The iPod and iTunes didn't just change the way we as consumers listen and store music; it also revolutionized the music industry.

iTunes was introduced to consumers in January 2001 and was originally marketed as a program that could convert CDs into Compressed Digital

Audio files and also organize digital recordings. It was later that year that Apple took its next major step and released the Apple iPod. This was a small storage device that contained an internal hard drive which could be utilized to store music files. The first iPod that Apple released had a 5GB capacity which averaged out to about 1,000 songs it could store.

Over the last fifteen years, both the iPod and iTunes have had a number of huge enhancements and new features released. All the new benefits and features have kept consumers coming back to the Apple brand, time and time again. It was in 2003 that Apple released the ability for consumers to be able to download individual songs for 99 cents. Every subsequent model of iPod has become smaller and sleeker, more stylish, and with larger internal storage and features than any previous model.

The next step in Apple's conquest was to release the Touch sensitive interface on the iPod. This unique touch sensitive technology gave users the ability to access their songs, movies, and the internet through Wi-Fi with a simple swipe of their fingers.

The iPhone by Apple

The next thing on our list was another product from Apple, the iPhone, which was released in June 2007. Love them or hate them, the Apple iPhone made a huge impact not only on the way we listened to music but on how we used our smartphones.

The original older versions of mobile phones were basically for voice calls and very basic text messages. Most of these older mobile phones had a keypad and either folded open or were considerably larger than the smartphones available today. Any phones that offered access to the internet were often extremely slow and hard to read. With the introduction of the iPhone, we saw a smooth multi-touch interface, which allowed users quick and easy access to the internet. Finally, users had a usable interface which allowed mobile access to the internet, almost anywhere around the world.

Apple also introduced apps; these were small, downloadable programs that users could download from the app store. This allowed Apple phone users to create their own custom interfaces on their phones to create unique experiences.

The Electric Car

The Tesla motor car company, Tesla Motors, was created in 2003 and released its first production model electric car in 2008. There have been a number of electric cars released since as early as the 1920s, but Tesla was the first company to release a production vehicle that had no major obstacles to prevent sales or production.

Most traditional petrol-powered vehicles produce an economy rating of approximately 20-25%, but the Tesla boasts an impressive 88%. Tesla has several different models in production, including a bot, a roadster, and also has plans to release a coupe.

YouTube

Unveiled in 2005, YouTube has had a huge impact on both the way we distribute media and also social media. YouTube uses Adobe Flash Video technology and provides a platform for users to upload and share their own videos. YouTube now allows users to create channels and profiles, share and promote their videos, and also make money through advertising opportunities on the site. Many celebrities have used YouTube as a jumping off platform to launch their careers, establishing a large fan base and following before moving on to mainstream media.

Teleportation

At the Australian National University in 2002, a successful research experiment resulted in a laser beam being transported. This research was the continuation of a previous experiment at Caltech in which they successfully transported a proton. The primary process involved using light and matter, with one serving as a carrier for the data and one was the storage medium. During the Australian National University experiment, the object was transported almost 1.6 meters.

The technology and its application have many, many possible uses. Through continued research, scientists are beginning to believe that teleportation of a variety of large objects over any amount of distance will become practical.

The AbioCor Artificial Heart

The first time that an AbioCor artificial heart was implanted into a patient was in the year 2001. This device was the first artificial heart that was completely self-contained and required no external batteries, wires, or any tubes. The AbioCor artificial heart utilizes an internal battery that is recharged by using a transcutaneous energy transmission process. The first model that was initially designed was limited by how long it would last, but the future models are expected to last a minimum of five years before they will need to be replaced. Scientists and developers ultimate goal are to design and artificial heart that lasts much longer than the life expectancy of the patient.

Blu-Ray Players

The Blu-ray Disc and the Blu-ray player was the next advancement in digital discs such as the DVD. Blu-ray discs and players allow much more data to be used than a DVD, and their primary use is for high-definition videos. Blu-ray was first introduced in 2006 and utilizes a blue laser to read and write data to the discs. The blue lasers utilize a much smaller profile wavelength than the traditional red lasers used on DVDs. The spread of Blu-rays has allowed high-definition video players and televisions to reach their full potential.

Facebook

Facebook wasn't the first social media networking website, but since its introduction in 2004, it has dominated the global scene. Over 150 million users around the world log in and take advantage of Facebook's features to stay up to date with friends, family, events, and celebrities around the world. Facebook is growing every year, introducing paid adverts and also suggestive ads based on each user's interests.

Solar Shingles

Solar shingles were introduced in 2005 by the Dow Chemical company. They developed a thin film of solar photovoltaic roof shingle material. These small-profile roof shingles can be easily and quickly integrated into any traditional asphalt shingles. The solar shingles cost almost 10-15% less than the installation of traditional solar panels. The Dow Chemical solar shingles have proven much quicker and easier to install than many other forms of solar panels.

Smart Bullets

Bullets have come a long way from a rough, round lead slug that was tamped down into place and usually resulted in the operator turning their head before firing. The first bullets and muskets relied heavily on volleys of fire at close range, with the speed of reloading determining the outcome of many exchanges. Smart bullets utilize a microchip which allows the bullet to detonate beyond an obstruction. Soldiers fighting in urban situations can use smart bullet technology to maximize accuracy and minimize civilian casualties.

Bluetooth Technology

We first saw Bluetooth technology hit the markets early in 2002, and it hasn't looked back since. Bluetooth was explained as the first low-energy peer-to-peer wireless technology. Bluetooth was developed by a large consortium of electronic manufacturers who were looking to connect several different digital devices over a short distance. The idea was first born inside an Ericsson lab in Sweden in the 1990s and has since gone on to revolutionize the wireless connection industry. The great thing about Bluetooth technology was that it allowed users with different brands and types of devices to connect to each other.

The first devices that utilized Bluetooth technology to hit the market were earpieces that enabled people to utilize hands-free technology on their phones. Now, Bluetooth technology is being utilized in a massive amount of different devices, smartphones, fitness devices, sports tracking, and many others.

Innovation, What Exactly Is It?

Innovation is defined as a new idea, or more effective process or device. The easiest way to view innovation is like a better solution or application which will meet new requirements or better meet existing market requirements. Innovation is accomplished through the use of more effective processes, products, technologies, services, or business models. An item can be great, but through innovation, it can be re-released into a market as a better product or service.

A device that is a new version of an already existing idea is often considered innovative, but in management science, economics, and other fields of analysis and practice, innovation is often considered to be a joint process. This process combines many different ideas in a productive way that have a positive impact on society.

The Difference Between Innovation and Invention

Many countries have a certain amount of money they put into budgets and set aside for innovation and innovative projects, but too often these projects fail to reach their full potential. One of the reasons for this is that too often innovation is confused with invention. If you go to all the effort to invent something, but then you don't do anything with this invention, then your efforts and your invention are wasted.

Innovation will take that invention, further its development, tweak it, improve it, and then market that invention and release it to the general population. A good example of this is Google and keyword ads. Google didn't invent keyword ads, but they did develop them to the point that they were practical and now play a vital role in how Google ranks pages on the Internet.

Modern Tools for Invention and Innovation

Innovation in Design Methods and Tools

With all great ideas and initiatives, innovation works best when it is performed in a project situation. As most people know, the best way to approach any project is with an organized team approach. This way, innovation is more likely to succeed and, with the correct subculture, future innovation will be more likely to grow in a practical environment. If your business or company promotes ideation, transparent values, initiative, and a rapid implementation of ideas, then it is more likely to have a better culture of growth.

All the different stages in innovation are aided by different tools and techniques. These all work together to get you towards the result, achieving valuable outputs through the exploitation of new ideas.

The Internet

The Internet is responsible for a very significant portion of the global GDP, or Gross Domestic Product. If you were to measure the expenditure and consumption of Internet-related purchases as an individual sector, it would now be larger than either energy or agriculture.

An excellent example of how the Internet was used to combine inspiration and innovation is in Togo, where a young engineer named Kodjo Afate Gnikou built a 3D printer. He surveyed a large amount of different electronic waste being deposited across West Africa. Using all of these miscellaneous electronic parts, he not only designed, but he also built a 3D printer. The electronic parts that he used were all from old computers, printers, and scanners. The combined total worth for all of these discarded parts was as little as $100. Kodjo named his 3D printer the W. Afate; it combined the name of where he worked at WoeLab's with his own name.

This, in many different ways, is just a normal story of how you can turn inspiration into innovation. This is the perfect example of how just one man can see opportunity where many other people just saw challenges. He used his intelligence and information from the Internet to create something that others just saw as worthless junk. Many inventions and innovations come from people seeing a challenge and accepting it, overcoming it, and providing a solution. The impressive thing about this invention was how Kodjo utilized information and services on the Internet to complete his printer. When you consider what he had to use and how he found his information, without the Internet this wouldn't have been possible 10-15 years ago.

Some of the different sources that Kodjo used to get the information he needed were:

- He utilized online collaboration between many enthusiasts on projects like the 3D printer, with plans made freely available for him to build the W. Afate printer.

- His next move was to raise funds for his project. He did this by reaching out to a global audience through a crowdfunding site call Ulule. From 112 supporters who had faith in his project, he managed to raise 4,313 euros.

- He also used global markets to submit his 3D printer for acknowledgment. The W. Afate 3D printer was sent to the NASA International Space Apps Challenge. The prize for this challenge was to have these applications possibly considered for any future missions to Mars aboard the space shuttles.

Most original high-tech projects or startups relied on producing physical products such as computers, peripherals, and chips, as well as the software that was needed to control these things. The creators and founders of these products and companies then saw the benefits from these innovations and inventions. The benefits were often excellent educations in the best universities, access to the latest technologies and computers, and also more investment to fund further inventions and innovations. These all combined in the early days to form around Silicon Valley, a place perfectly suited and created for nurturing startups with successful results.

By combining all of this information, it is no wonder that some of the greatest Internet startups of the 1990s—eBay, Yahoo!, Amazon, and Google—all originated not only in the United States of America, but in Silicon Valley. The thing that all of these companies had in common was that they all flourished in the conditions that many earlier tech startups had experienced. They also all came at a time when the United States had the early lead in access and usage of the Internet.

Now that the Internet is available on a much wider scale, it has changed the opportunities for startups and innovations around the world considerably. No longer do startups have to be located in Silicon Valley or the United States to receive the same benefits that they used to in the early 1990s. We no longer have to have any sort of physical presence to access the ingredient we need to build our products, and everything can be organized remotely, from development all the way through to sales. As the Internet spread its way across the developed world, we began to see new startups popping up. Two of the most well-known startups that began their lives outside of the United States in the 2000s were Skype, which began in Estonia, and Spotify, which began in Sweden.

Recently we have seen that the center of the Internet has been shifting from the developed world to the developing world. Figures show that in 2008 the numbers of people accessing the Internet in developing countries passed those accessing and using the Internet in developed countries.

One thing that we touched on in previous chapters was the iPhone. The iPhone was first launched by Apple in 2007 and by late in 2011 more than 50% of its users had access to mobile broadband. When you combine this with the first trend, it shows that by the end of 2012 more than 50% of world's mobile broadband was being accessed by users in developing countries.

When you combine both of these trends, it points in one direction: that future innovations, inventions, and developments involving the Internet are more likely to come from developing countries.

Some of the startups we see already come from developing countries. One of these is M-Pesa, which is arguably one of the most widely used mobile payment systems found throughout the world and which began its life in Kenya. It is no surprise that M-Pesa is doing so well since banking and mobile payment systems are in high demand in many developing countries. Another successful application that has come out of Kenya was Ushahidi, which was originally developed to meet a local need. This application was initially used to track violence after a local election which was held in 2007. Since then the application has been adopted for use in a variety of different places such as Haiti and Japan for earthquake relief operations and also in Washington DC to track snow plows.

It is important that the Internet continues to be governed not by one group or agency but by a multi-stakeholder type of participation. This will allow anyone to utilize inspiration and turn this into innovation and that innovation into the invention.

3D Printing and Invention

We have already discussed in the previous section how using materials that otherwise might be trash and with plans freely available on the internet you can build a functional 3D printer. It may not be the best model available and not up to the requirements of some of the things we are going to talk about next, but it would definitely work.

If you haven't ever seen it done then, 3D printing can sound pretty damn cool, like something straight out of the Jetsons television show. Watching something appear before your eyes with just a few clicks of a keyboard or a mouse is amazing. Realistically, the technology itself is pretty straightforward and easy to understand; instead of printing with ink you use a substance that will layer upon itself like a plastic resin. It may sound simple and look simple but, in reality, it is quite an amazing leap forward for humanity.

As the technology itself spreads around the world, and the development of the machines and technology becomes cheaper, the biggest implication will be on the manufacturing industry. Goods won't need to be made thousands of miles away; they could be printed out down the road, factories full of 3D printers printing everything that we need. This could even lead towards people having their own 3D printer in their homes and being able to print out individual items, drastically reducing the cost of the items. Initially, the consumer would need to buy the 3D printer, the software, and the consumables, but after the initial purchases, only the consumable material used to manufacture the goods would need to be bought again. Even if it actually cost slightly more to make the item, the savings from transportation, shipping, and going to pick up the item would balance it out or make it cheaper.

If you look at it this way, we'll use cars as an example. Cars used to be manufactured locally in large factories and then shipped around the country. Now, many car manufacturers have moved their manufacturing plants overseas, where the costs of labor and raw materials are cheaper. We could see this trend reversed with the advancement of 3D printers. Parts could be assembled locally; dealerships could produce spare parts on request at their locations and assembly plants could be located locally, eliminating the need for supply chains and shipping.

The other side of 3D printing is that goods or consumables will be much more customized. With a simple press of the button, specifications can be changed, and the items won't need to be customized like they once were. Once you produce an item once, you can simply switch between programs whenever you like and reproduce that item without having to change machine settings or operations.

As 3D printing advances many of the reasons for companies producing their goods in countries such as Vietnam, China or Korea will no longer be valid. Countries such as China may see a significant drop in their manufacturing and export for a limited time, but it won't kill it altogether. As with other technological advancements, they will need to change and adapt to meet new market requirements.

China has positioned itself as a major player in the manufacturing industry over the last few decades, securing manufacturing from most of the strong economies around the world, by pushing their mass manufacturing to the very limit. By producing massive amounts of consumable products, it pushes prices down and also minimizes the most expensive aspect of any production, which is labor. The Chinese government isn't shy about just how pro-producer they are, favoring boosting the country's growth over the living standards and purchasing power of its own citizens. The problem with this is that as 3D printing advances and countries begin utilizing its technology to manufacture goods, China won't be able to reduce already bare minimum wages to offset the costs of no shipping.

Many Western countries like the United Kingdom, the USA, and Australia that import a lot of their small goods may see a revival in manufacturing. China won't lose it all; some products just won't be suited to 3D printing, and it has a large local population to produce for, but it may very well take a substantial hit. We have seen a lot of manufacturing power shipped over to the East over the last few decades; it may be time to see a lot of it returned, but only the future will tell how 3D printing will really implicate manufacturing.

12 Cool and Exciting Things That 3D Printing Is Doing Right Now

If you wanted to right now and had the money, you could pay $850,000 approximately, buy yourself a 3D printer, and the materials needed to print yourself your own guns at home. Sounds pretty scary? It would also be completely legal. Now isn't really time to worry about that, though, and if you had over $850,000 to spend on guns you could certainly get yourself a much bigger weapon than whatever you could print at home. What we do need to focus on, though, is the fact that 3D printers are here, and people are creating some truly life-changing items with them. 3D printing is bringing in an entirely new generation of inventions and ideas.

Over the last few years, many engineers have been utilizing 3D printers to help them create 3D scale models of parts and prototypes. These were basic 3D printers that utilized a thin plastic filament to produce these small models, but lately, 3D printing has really taken off. 3D printers are becoming much better and, most all, they are becoming cheaper. As they get more advanced, and their use spreads, the models themselves become cheaper and much more achievable for more people. What once was used to print a small piece or prototype is now being used to print out complete car parts or even body parts. You can already buy small 3D printers for kids to use at home to make a wide variety of different things.

The following things are just some of the amazing things that are being produced using 3D printers around the world right now.

Prosthetics

At the London 3D Printshow a design firm, Fripp Design & Research, unveiled their very realistic, life-like prostheses. The conventional manufacturing process for most prosthetics is extremely expensive. As an example, a silicone ear or nose may cost as much as $4,000 each. An impression is taken from the injured area, and then a sculpting is made out of wax, finishing any small details, and then the final shape is made out of silicon.

Using a 3D printer, manufacturers can take a photo using a digital camera, capture the image of the affected area, and use the patient's actual skin tone to create a digital model of the part that is required. The final digital image is then sent to a Z Corp Z510 color 3D printer and the parts are then printed. The initial one off part is approximately the same price as a traditional piece but, as they quite often need replacement over time, any replacement parts are only approximately $150.

Bones

Recently, researchers at the Washington State University have been able to use 3D printers to print off replacement bones for certain patients that required orthopedic or dental procedures. The pieces that they made were used as bridges, placed surgically next to bone and then acted as support while the bone healed itself. As the bone was healed the 3D printed pieces then dissolved, leaving behind only the organic healed nose, for instance.

Pizza

In September at Austin's SXSW Eco, NASA debuted a pizza printer. They are still working on creating the perfect recipe! The pizza they cooked during the expo was a basic dough, topped with a ketchup and cream cheese. It was like pizza, but not quite like a pizza! A heated plate is then used to cook the pizza as it is printed.

NASA has shown a lot of interest in 3D printers and what their applications could mean for space travel. If you could send one 3D printer into space along with all the dehydrated ingredients necessary it would eliminate the need to send much larger and bulkier prepared foods. For long distance space travel, they are looking to achieve at least a 15-year shelf life with the foods they send. Using dehydrated materials and the 3D printer, they could expand this out to almost 30 years. Imagine the first people to ever foot set on Mars doing so with a slice of freshly cooked pizza, straight out of the 3D printer.

Desserts

Okay, we all love desserts, we just don't love what they do for our diets and our waistlines! So, imagine that you feel like a donut, but you know that they are bad, so instead of just grabbing any old diet your diet is printed out for you, containing the exactly right amount of fiber, calories, and protein suited to your diet. Now your donut isn't just a naughty snack; it is a naughty snack that also has a healthy amount of fiber and extra vitamins if you needed them. Now, imagine that this donut isn't being cooked at a donut shop, it is being printed out for you by a 3D printer.

Researchers at the Cornel University's Fab@Home lab have demonstrated a 3D printer that did something very much like this, but they printed cookies. Two test subjects put all of their information into the computer, weight, height, and BMI. The 3D printer then analyzed all of this information and, based off of that information, printed out two snowflake-shaped cookies that to all appearances looked identical, but they weren't. The 3D printer had analyzed each person's statistics and printed them cookies to suit their dietary needs and requirements.

Clothes

It doesn't take long for large corporations to take advantage of any advancements in technology, and 3D printing is no different. Nike has begun printing a football cleat which they have named the Vapor Laser Talon and it is supposedly able to help athletes run faster. One of their rivals, New Balance, is also using 3D technology. New Balance is using technology to scan their customers' feet and then use 3D printers to print out shoes made specifically for them, made using nylon polymer.

To the fashion industry, customization and personalization have long been the essential ingredients to haute couture. Much like the rest of the world, 3D printing has begun a slow invasion of the fashion industry. Architect Bradley Rothenberg used 3D technology to scan Victoria's Secret models and then gave each of them a one-of-a-kind custom set of angel wings to walk in during the lingerie show. Another designer, Iris van Herpen, a Dutch designer, also showed off a set of 3D printed shoes at Paris Fashion Week.

Another designer, Janne Kyttanen, used 3D technology to design a wedge heel that could be printed off ready to wear the next day.

Guitars

When you design guitars you have a lot more freedom of design compared to, say, a violin, whose shape and sound never change much from the classical standard. At last year's Maker Faire exhibition, a fully-functional Gibson Les Paul guitar was printed using a 3D printer.

If you prefer your guitars with more of an acoustic feel, then it is also possible to print acoustic guitars. A designer, Scott Summit, spent his entire childhood just lusting after a $3,000 guitar exactly like Jerry Garcia's. He never could afford to buy it as a child, so he did the next best thing: he printed one using a 3D printer. Apparently he never expected it to sound great, but it sounded a lot better than he expected it to. Summit expects that in the future musicians will be able to go online and order specific custom guitars which would be able to produce custom, unique sounds.

Fetuses

If you or someone you know has ever been pregnant and, let's face it, everyone knows someone who is having a baby, then you have definitely seen an ultrasound photo. You may already be sick of seeing every possible blurry and hard to distinguish ultrasound picture. Now, a mother can simply have her fetus printed out in a life-size 3D printed model.

Fasotec, a Japanese company that specializes in 3D printing, combined with Tokyo's Hiroo Ladies Clinic, has begun a service that produces an actual life-size replica of a woman's abdomen. Don't worry, though; the fetus is fine! A copy of the fetus is printed using a clear filament, and then the fetus is printed using a white filament so that it stands out. If you love your baby, and you love baby mementos, then you are definitely going to need a 3D printed baby fetus for your coffee table!

Stem Cells

If you are against animal testing and, let's face it, most of the world is, then you will be happy to know that now there is a new alternative. Alan Faulkner-Jones demonstrated some revolutionary new technology recently at the London 3D Printshow. Alan Faulkner-Jones, a Heriot-Watt University researcher, modified a 3D printer to print out micro-tissues and micro-organs that could in four or five years be used to effectively test prescription drugs. Testing on animals like rats, monkeys, or bunnies could be a thing of the past.

Alan Faulkner-Jones said that his technology could be taken even further. A patient could have their specific cells printed off and then these cells would be used to test responses to specific medication before it has been administered. This way, if there were any adverse reactions, doctors would be aware of it before they even administered the medication.

Chocolate

If we have ever done anything right as a species then it could be this. Yes, we have perfected the art of printing 3D chocolate. Imagine sitting back in the comfort of your own home and the sweet tooth starts crying for a sweet. You don't want to go out, so what do you do? Oh yes, you just print out some delicious chocolate on your handy 3D printer! Well, we may not be quite at that stage yet, but it isn't far away.

Artists in South Africa have used a modified a MakerBot 3D printer to print chocolate sculptures. They didn't just do it for fun; the project was actually part of a campaign to promote Android's new KitKat operating system that it is rolling out. The entire collection of printed chocolate sculptures was displayed at the Museum of African Design in Johannesburg for several days as part of an entirely chocolate-themed exhibition called Chronology.

Cars

Mechanics have been using 3D printers to do automotive parts for years, and so has Jay Leno. This same technology could be especially handy for soldiers or explorers who are miles away from their forward supply bases. If a part of a machine or weapon or any other piece of vital equipment broke, then they could simply reproduce it using their 3D printers. Obviously, they would need to have power of some sort and the raw materials, but a generator and the raw material wouldn't be as much to carry around compared to one or two spare parts for every piece of machinery or equipment.

Any time now, two twenty-year-old brothers are planning on setting a new record. They are going to drive a 3D-printed car across the United States of America. They have named their pod-shaped car the Urbee 2. They have designed the car to be as economical and environmentally friendly as possible, hoping to get an excellent 300 miles per gallon of fuel. The brothers, Cody and Tyler Kor, in addition to their dog, are planning on reproducing the route taken by the first two men who ever drove across America in a motor vehicle.

Cats

If you love your cat more than you love your children, then if anything ever happens now you can just run out and print another one! Well, not exactly! Not yet anyway, but it would be kind of cool and weird all rolled together.

The MakerBot 3D printer already has plans to print cat statues; not actual cats, but just their statues. You can go online and use their downloadable plans. Just choose the model of cat you would like, select certain features and fur color, and you are ready to go. Now if your most precious feline friend departs this world early, you can simply choose an extremely life-like model, print it out, and you will always have your companion around to keep you company.

3D printing is advancing at an extremely rapid pace all throughout the word. 3D printers are helping inventors create parts and prototypes where before expensive, and time-consuming, parts would need to be produced by hand in most cases. Now, if you have an idea and have access to a 3D printer, you can draw out your parts using a software program, design your invention, and print out all the parts or the entire prototype.

Inventions in the 21st century are going to advance at a much faster pace thanks in part to one particular invention, the 3D printer.

5. CROWDSOURCING

The History Behind Crowdsourcing

Crowdsourcing isn't a new idea; weird, right? Everyone believes that this is some modern idea that was recently invented, but they would be wrong! Crowdsourcing as an idea has been around for quite a long time. The term "crowdsourcing" was a relatively new idea; it was invented by the editors of Wired magazine, Jeff Howe and Mark Robinson, in 2005. It wasn't until the next year when the term was used in an article and blog that the term itself started to take off and gain momentum and be generally known around the world. New name, same great idea from centuries earlier, maybe with a small twist.

New name, same great idea, crowdsourcing is basically taking a large project and then breaking it into smaller pieces, making it easier to accomplish. It allows one person to organize a group of people, the person or people at the center of the organization give each part of their project to someone else, and they all work independently of each other. As they finish their small piece of the puzzle, they then add it to the other people's until the entire project is then assembled and completed. For large or complicated projects this basic idea works quite well.

This idea of small, independent people or groups working together on one project only really started to gain momentum when communication and travel became much easier. People can talk to each other all around the world, anytime, almost any place in the world they like. The Internet has made working online and communication much easier. You can order items online from almost anywhere in the world or have things delivered to anyone in the world almost overnight. Times certainly have changed, and all of these positive changes have made it much easier for remote projects to be worked on together.

Before the world really entered the digital era that we are all currently living in and enjoy, when you had a job you showed up to the place you were employed, and that's where you worked. If you created software, parts, cars, or any other item, everyone basically showed up and worked on their products at their workplace. Through the use of modern travel and communication, project managers can now employ people from anywhere in the world. Places like Fiverr, Upwork, Guru, and many others all make this possible.

Innovation Through Using a Crowd

There is one pretty famous example of invention and innovation through the use of a crowd or group of people, and you might be surprised to know that it was three hundred years ago. It was in 1714 that the British government turned to the general public to get their help in solving a problem, one that they were having trouble solving alone. In 1714, the British government offered a reward of £20,000, which is equal to about $3.5 million today, for anyone that could develop a method for ships to determine their longitude while they were at school.

There were a lot of interested people but, in the end, there were two main groups that were closing in on a solution to the problem. The first was a team of professional astronomers and the second was just an ordinary, working class clockmaker. Both technically solved the problem, each working towards a solution from different angles. Unfortunately for both of them, the British government refused to pay the promised reward on a technicality. Today the entire world would know within hours if the same thing was to occur, lawyers would become involved, the whole world would soon know that the government had failed to uphold their side of the bargain. Eventually, that clock worker's invention would become the foundation for the world's first marine chronometer.

Despite the fact that they missed out on receiving the full prize that they thought that they were entitled to, the entire project proved that the idea of an invention or innovation could be accomplished through the use of an incentive. Now that we have the Internet, companies from all around the world can come up with challenges or think-tanks where they can tap into some of the world's brightest and smartest. The Internet is drawing people from all walks of life into one large community where ideas can grow and mature, with no restrictions on where you live, who you are, or what religion it is that you practice.

An example of a modern version of the British government's original idea for a prize-based solution to a problem can be seen with the virtual solutions generator InnoCentive. They post a variety of different science and technology challenges online through their website. These challenges are open to anyone but focus mainly on scientists, students, and engineers from around the world. If you submit a solution to one of the problems, and it is proven to be one of the better solutions, then you are likely to win cash prizes and the honor of having provided a solution and the prestige amongst your peers. There is also the bonus of having successfully helped to achieve a solution to a problem that could ultimately benefit mankind.

Now that we have hit the 21st century, we are seeing companies really beginning to take advantage of the idea of having their work completed in a distributed manner. We have all had computers in our homes for the last ten to twenty years, but only during the last few years have they become affordable for everyone and the Internet makes the world much more accessible to everyone, no matter where you are or where you live. The Internet is available on your smartphone through satellites pretty much everywhere nowadays.

The author James Surowiecki published a book in 2004 titled The Wisdom of Crowds, a book entirely about the acts of distributing intellectual ideas and creative tasks to a multitude of different people to get a large amount of return. He argued in this book that through the use of crowds you could get a better result than through the use of individual skilled workers. The book focuses on how, through the use of a large diversified group, you could get a better result than that of using skilled individuals. The key, he claimed, to the success of crowdsourcing was a technique that he called aggregation. This system is how you utilize a group's individual efforts and then combine them to get a positive result.

When you combine both the 19th century idea with the technology from the 21st century you get an idea that works and not only just works, but one that works amazingly well! For example, if you want to release a book but you don't have the time to write it yourself, there is a solution. You could go to a freelance site like Fiverr or Upwork, hire a writer to write your basic outline and they can then work through and write the entire book. Now you hire a proofreader to proof and edit your book, then a formatter to do the layout of the book. Next would be a graphic designer to create your covers, and finally an eBook specialist to choose categories and keywords for publishing the book. In a matter of weeks or a month, you have an entire book ready to publish, all made possible through crowdsourcing and freelancers from around the world!

In 2015, a large survey conducted in the United States of America found that roughly 54 million people had tried or were doing some sort of freelancing work online. Another popular example of crowdsourcing is Wikipedia; this online encyclopedia was built and is managed by thousands of volunteers everywhere around the world. Even the United States Department of Defense has used crowdsourcing to try and find solutions to disaster management problems.

Crowdsourcing has been around for centuries, but it wasn't until these recent advancements in technology, like computers and the Internet really took off, that crowdsourcing became a truly viable solution to businesses problems.

What Is Co-creation and How Are Companies Using It?

Co-creation utilizes the insights of your consumers and then gathers this information and input and points it towards innovation of your product or service. As consumers become more active in interacting with and shaping the products that they buy, CEOs of large corporations are more and more interested in gathering this data and taking advantage of it to drive innovation. Once upon a time, companies created products, released these products, and then gathered the feedback from their consumers. Now, we are seeing consumers interacting with the large corporation through social media and helping to shape and drive innovation.

In the following section, we are going to talk about how some of the world's largest corporations are utilizing co-creation to drive innovation in their products and their services. For years, corporations have been using their consumers to drive their innovation, but only recently has it really taken off. This spike is largely due to that fact that consumers now have much more access to the Internet, personal computers, and smartphones.

The first company we are going to speak about is Coca-Cola and how they have been using co-creation drive innovation throughout their global empire. There aren't many places in the world where you can't buy a version of Coca-Cola. One clever example of how Coca-Cola is using co-creation is through their FreeStyle Fountain machine available in a wide variety of different fast food outlets. You have probably used one yourself; you get to choose which flavors you want to mix and then create your own flavors. Instead of them placing ten drinks in a fountain machine, they place hundreds of possibilities and then allow consumers to create their own favorite drinks. They can then gather the information about the drinks created, favorite flavor combinations, and then use all of this consumer-generated information to create individual flavors for mass production. They then combine this new FreeStyle fountain machine with a new mobile app for smartphones. This mobile app allows consumers to save their favorite blends so that no matter which FreeStyle machine they use anywhere in the world they can load up their own creations.

We are now going to change directions slightly and look at another huge corporation, one that deals in software, not soft drinks. They work with their consumers to create ads that not only work, but ads that also work better than their old formats. They understand that no one really wants to see ads, but because advertisements are an integral part of gathering revenue, they want to see ads that are relevant to what they are interested in.

LEGO has been around for a long time, they are well accustomed to getting into the heads of young children and learning what it is that inspires them, but they don't always get it right. When sales were starting to slow down in the late 1990s, LEGO started looking around at finding a solution to their problem. They decided that kids were more interested in characters that were less buildable, like an action figure. They released an action figure-type character in 2001 called Jack Stone. There was only one problem: kids don't care about an action figure without a back story or any history. They love action figures from movies because they already understand the story and can relate to it. In one quick move, LEGO alienated all of its base consumers by switching from the traditional building bricks that it had built its company on.

Large corporations are only just beginning to tap into co-creation, and the future is going to be a bright place with much more interaction between consumers and corporations.

What Is Co-design and How Are Companies Using It?

A co-design workshop can be a fantastic way for large corporations or companies to get together with consumers and designers to gather their perspective when coming up with new ideas and innovation for existing products. When consumers feel as if they are a key part of the design process of products, then they are more likely to not only support the product but relate to it.

Through the use of co-design, companies can allow users to become an active part of the process of designing and testing new products or services. The co-design mentality is based on the fact that, no matter where someone is from or how educated they are, they all have something that can be contributed to a project. Everyone brings their own unique opinions and points of view to any project. The principals of co-design can be used at any stage of a project, but at the beginning of any new product development or project is the ideal time. You may believe that you have looked at a project or product from every possible angle, but you put that product or service in front of 1,000 people that have no vested interest, and they will quickly help you identify any pros or cons. Ultimately you are developing a product or service which you hope will benefit your customers, so getting their input and ideas can be priceless at the early stages of the development.

The following sections help you identify the different stages of co-design and how you could implement them yourself on any of your own projects or products.

Self-Reflection Research Methods

A healthy co-design service can end up giving you a large amount of very useful data. The amount of data and what type of data will vary depending on what product or project that you are working on, product maps, service plans and rollout ideas, product development, or innovations to products and services.

You need to work out what you are hoping to achieve from your co-design sessions and then develop a group of people who understand the goals and can work with you to help identify the things you are looking for. Once you get the results of these initial sessions, then you have a good idea of where to continue working on. You need to work out what it is you want and then tailor the sessions to achieving these results, but not so rigid that you rule out the possibility of learning other ideas. You want your sessions to unveil new data, but not be so constrictive that you hinder the process.

Running Your Co-Design Workshops Onsite

You can run any co-design workshops either in a specific location, or if the service or product doesn't need to be physically used then, you could set up an online testing workshop. You need to structure the way that you run any co-design sessions carefully; you are trying to achieve certain results. You want to guide the participants towards the way you feel will give your product or service the best outcomes for your research. It isn't any good to you if you have 1,000 people testing a new program or product, but if 999 of them get wrapped up in a certain part of it and never use the part you want them to, then the entire process will be a waste of time. You need to set them goals or tasks to keep them on track, have them fill out questionnaires, answer questions, and do certain tasks that will help you gather the information that you need.

If you are going to be working with a large group of people in one location, then it is important that you have enough representatives to help guide and answer any questions. If you are conducting your co-design sessions online then this will be much easier to organize, but if you are doing it at one of your locations, then you may wish to separate large groups into much smaller groups. This serves two purposes: firstly, it allows people to speak freely and form their own opinions and then express them in smaller groups. Secondly, it will allow you to assign your representatives to smaller groups and then make it easier to answer questions or guide people towards the direction you want them to go in the session.

You want to allow enough time for everyone to experience the product or service but not become bored with it. If your product is simple, then don't expect people to be able to spend hours investigating it. If it is more complicated, then it is going to need more time. Most group co-design sessions normally run between one or two hours, if you run them too long then they will lose interest, if they lose interest, then you could lose valuable information. Try to keep the people motivated and enjoying themselves, gather their thoughts throughout the process, don't wait until the end to ask their thoughts. If you feel as if you can organize it, then break up the sessions to encourage people to provide you with their thoughts.

Remember, you are trying to gather people's base thoughts, so you don't want them to be telling you what they think you want to hear … you want them to tell you what it is they are truly thinking and feeling about the product or service. If you can gather this information in the most natural forum, then you are going to get yourself information gold!

Hopefully, by using this information, researchers, stakeholders, designers, and engineers can create products that are better suited to their consumers.

ABOUT THE AUTHOR

Poorya Montaseri

Since early childhood, curious Poorya had a knack for breaking things and understanding the inner working of complex machines. As he grew up he put his creativity and innovative ideas into work by bringing out-of-the-box thinking to everyday business practices while keeping his inventive mind busy with side projects.

As a serial entrepreneur, Poorya made his mark by founding an award-winning event management company bringing leading edge technologies to the industry. As an experienced businessman and certified life coach he has been helping hundreds of aspiring entrepreneurs transform their businesses and their lives. Poorya serves the community by raising awareness and educating people from all walks of life on creativity, innovation, and business.

As a part of his creative meditations, he often works on ideas that could address some of the major challenges people face on daily basis. GANGO (patent pending), is one of those ideas. By observing seemingly trivial events in everyday life, Poorya was able to find a solution that could revolutionize how vehicles are washed conveniently and efficiently.

In 2016, Poorya has received a number of accolades for his latest invention:

- Bronze Medal, Geneva Invention Convention, March 2016
- Thailand Award for International Invention, March 2016, and
- Croatia Award for International Invention, March 2016.

Get in touch with Poorya on:

- GANGO Website: http://gango.io, and

- LinkedIn: https://linkedin.com/in/poorya-montaseri-31176656.

ABOUT THE EDITOR

Ali Shabdar

Growing up in the early 80's, Ali spent most of his time tinkering with electronics and building circuits form scratch. During high school years he fell in love with computer programming, an endeavor that ultimately turned into a career for him.

Throughout the course of his career, Ali has created software for many organizations, trained a wide variety of audiences on technology and business, and published his first tech book in 2009. Many years later, his best friend Poorya shared an idea of a new invention with him. He got excited and jumped on board with no hesitation. Today, Ali works closely with Poorya on a number of inventive and innovative ideas.

Get in touch with Ali on:

- Twitter: @shabdar, and

- LinkedIn: https://linkedin.com/in/shabdar.

REFERENCES

Human Evolution: The Origin of Tool Use. (n.d.). Retrieved August 31, 2016, from http://www.livescience.com/7968-human-evolution-origin-tool.html

Sample, I. (2012). Lethal weapons may have given early humans edge over Neanderthals. Retrieved August 31, 2016, from https://www.theguardian.com/science/2012/nov/07/lethal-weapons-early-humans-neanderthals

When Did People Begin Using and Controlling Fire? (n.d.). Retrieved August 31, 2016, from http://archaeology.about.com/od/ancientdailylife/qt/fire_control_2.htm

History and Evolution. (n.d.). Retrieved August 31, 2016, from http://sheltertwc.weebly.com/history-and-evolution.html

Humans First Wore Clothing 170,000 Years Ago. Retrieved August 31, 2016, from http://www.seeker.com/humans-first-wore-clothing-170000-years-ago-1765156178.html

Smithsonian's National Museum of Natural History. (n.d.). Social Life | The Smithsonian Institution's Human Origins Program. Retrieved September 01, 2016, from http://humanorigins.si.edu/human-characteristics/social

Timeline and Inventions of the Middle Ages. (n.d.). Retrieved September 01, 2016, from http://inventors.about.com/od/timelines/a/MiddleAges.htm

Top 10 greatest inventions of ancient China. (n.d.). Retrieved September 01, 2016, from http://www.china.org.cn/top10/2011-03/04/content_22054243.htm

Persia Cradle of Science, Technology. (n.d.). Retrieved September 01, 2016, from http://www.iranreview.org/content/Documents/Persia_Cradle_of_Science_Technology.htm

Lavin, T. (n.d.). 8 Great Modern Innovations We Can Thank Muslims For. Retrieved September 01, 2016, from http://www.huffingtonpost.com/2015/01/06/8-inventions-we-can-thank_n_6424836.html

The Television to the railway steam locomotive: Ten of the greatest British inventions. (2011). Retrieved September 01, 2016, from http://www.dailymail.co.uk/home/moslive/article-2034658/10-greatest-British-inventions-From-television-railway-steam-locomotive.html

10 Inventions You Had No Idea Were French. (2014). Retrieved September 01, 2016, from http://frenchmorning.com/en/10-inventions-idea-french/

Inventions and Inventors. (n.d.). Retrieved September 01, 2016, from http://www.inventionware.com/renaissance-inventions/

Leonardo da Vinci: An Inventor Ahead of His Time. (n.d.). Retrieved September 01, 2016, from http://www.da-vinci-inventions.com/

Top 10 Leonardo da Vinci Inventions. (2013). Retrieved September 01, 2016, from http://www.geniusstuff.com/blog/list/10-leonardo-da-vinci-inventions/

History.com Staff. (2009). Industrial Revolution. Retrieved September 01, 2016, from http://www.history.com/topics/industrial-revolution

Timeline of Inventions of the 20th Century. (n.d.). Retrieved September 01, 2016, from http://inventors.about.com/od/timelines/a/twentieth.htm

Impact of the Industrial Revolution | Ecology Global Network. (2011). Retrieved September 01, 2016, from http://www.ecology.com/2011/09/18/ecological-impact-industrial-revolution/

Top 10 Inventions of the 20th Century - Toptenz.net. (2010). Retrieved September 01, 2016, from http://www.toptenz.net/top-10-inventions-of-the-20th-century.php?utm_source=feedburner

Rogers, S. (2011). Nuclear power plant accidents: Listed and ranked since 1952. Retrieved September 01, 2016, from http://www.theguardian.com/news/datablog/2011/mar/14/nuclear-power-plant-accidents-list-rank

Best Inventions of the 21st Century: Read Our List of the Top 10. (n.d.). Retrieved September 01, 2016, from http://www.brighthub.com/office/entrepreneurs/articles/86314.aspx

25 Most Intriguing Inventions Of The 21st Century. (2015). Retrieved September 01, 2016, from http://list25.com/25-spectacular-inventions-of-the-21st-century/

21st century inventions that made an impact. (n.d.). Retrieved September 01, 2016, from http://www.inventions-handbook.com/21st-century-inventions.html

Innovation. (n.d.). Retrieved September 01, 2016, from https://en.wikipedia.org/wiki/Innovation

The difference between innovation and invention | ZDNet. (n.d.). Retrieved September 01, 2016, from http://www.zdnet.com/article/the-difference-between-innovation-and-invention/

Innovation Tools and Techniques. (n.d.). Retrieved September 01, 2016, from http://www.cse.dcu.ie/our_services/innovation/tools-and-tech.html

Internet matters: The Net's sweeping impact on growth, jobs, and prosperity. (n.d.). Retrieved September 01, 2016, from http://www.mckinsey.com/industries/high-tech/our-insights/internet-matters

From Inspiration to Innovation On the Open Internet. (n.d.). Retrieved September 01, 2016, from http://www.wired.com/insights/2014/06/inspiration-innovation-open-internet/

Boboltz, S. (n.d.). 11 Amazing Ways People Are Using 3D Printers For Good, Not Guns. Retrieved September 01, 2016, from http://www.huffingtonpost.com/2013/11/15/3d-printer-inventions_n_4262091.html

3-D Printing Will Change the World. (2013). Retrieved September 01, 2016, from https://hbr.org/2013/03/3-d-printing-will-change-the-world

Crowdsourcing: An Old Idea Amplified by Modern Technology - OneSpace. (2016). Retrieved September 01, 2016, from http://www.onespace.com/blog/2016/03/crowdsourcing-old-idea-amplified-by-technology/

Co-creation: 5 examples of brands driving customer-driven innovation. (2016). Retrieved September 01, 2016, from https://www.visioncritical.com/5-examples-how-brands-are-using-co-creation/?lb=1

Creativity-based Research: The Process of Co-Designing with Users. (n.d.). Retrieved September 01, 2016, from https://uxmag.com/articles/creativity-based-research-the-process-of-co-designing-with-users

www.ingramcontent.com/pod-product-compliance
Lightning Source LLC
Chambersburg PA
CBHW060415190526
45169CB00002B/912